green greener greenest

Façades
Roofs
Indoors

The Deutsche Nationalbibliothek lists this pub-
lication in the Deutsche Nationalbibliografie;
detailed bibliographic data are available on the
Internet at http://dnb.dnb.de

ISBN 978-3-03768-212-8
© 2017 by Braun Publishing AG
www.braun-publishing.ch

1st edition 2017

Editor: Editorial Office van Uffelen
Editorial staff and layout: Lisa Rogers,
Johanna Schröder, Nora Meyer,
Fanny Klang, Katrin Bühner
Translation: Judith Vonberg, London
Graphic concept: Michaela Prinz, Berlin
Reproduction: Bild1Druck GmbH, Berlin

Chris van Uffelen

green greener greenest

Façades
Roofs
Indoors

BRAUN

Contents

Indoors

6

Since the 1970s, an awareness of ecological issues has increasingly become an important ingredient in good architecture. What began as part of the "Jute Not Plastic" movement and competed openly with developments in high tech has since joined forces with its supposed opponent and embraced the new esthetic that has emerged. Neither the high tech of the Centre Pompidou in Paris nor the ecological approach of Peter Vetsch's Hobbit-like earth houses could defeat the other, and both have made a significant contribution to the functionality and esthetics of architecture, which today has ecology as a priority.

At first it was high tech architecture that took the leading role in the development of this symbiosis. Ingenious solutions to back-ventilated façades, ventilation systems, passive heat trapping and the use of renewable energies dramatically reduced the energy requirements of the structure being designed. Within a few decades, these techniques became an essential

Fig. 3.

part of any architectural project, but were for a while also stylistic elements that contributed to an image of technological-ecological design. After a relatively short period of time, it was deemed unappealing to have solar panels on display.

Yet, regardless of a feature's ecological value, interest in the appearance of such value remained strong. Through greened structures and natural materials, early examples of ecological architecture could thus begin to make their mark.

Façade greening

Façade greening is both the oldest and the most recent manifestation of greened architecture. The hanging gardens of Babylon are probably the wonder of the ancient world that is most shrouded in legend. Not only are there scarcely any written sources of information about the structure, but it was only in the modern era that the gardens were designated a "wonder" and their historical existence is surrounded by mystery. The terrace construction was supposedly situated directly next to or within the palace of Babylon. Nebuchadnezzar II created the "hanging gardens" around 600 BC, probably for his wife Amyitis. The first descriptions originate in the fourth century BC and it is assumed that the gardens were destroyed along with the city no later than a century before the birth of Christ. What is known for sure is that the phrase "hanging gardens" is a false translation from the Greek

– "greened roof terraces" would be more accurate. Nevertheless, green façades have existed for many centuries, integrating nature into architecture with climbing or hanging plants. Yet this greening usually took place only after the construction phase was completed and without the knowledge or even the consent of the architect. In these examples, the façade was not greened in the narrow sense of the word, but rather transformed with the addition of plants; the greening was not a component of the architecture but instead obscured it. Yet the ivy, climbing roses or vines fulfilled their purpose: they brought nature closer to those living in civilized environments or provided shade for the interior. The twenty-first century has seen a new form of façade greening, however. When Patrick Blanc created a vertical garden for the interior of Andrée Putman's Pershing Hall in Paris, a fresh kind of landscape architecture was born. Tropical plants, grasses and shrubs are planted on a wall and grow upwards. What was previously only possible to achieve on much smaller surfaces was now replicated on façades that were the height of buildings. A change in the Parisian building regulations in 2006 led to the city becoming a hub for this new kind of design. Thirty-nine vertical gardens were created in that year alone in the city. For each one, special irrigation systems were necessary and the supporting mechanism had to be affixed to the exterior wall in the same way as a façade cladding. For every greening, a detailed plant plan is required to ensure the façade is esthetically pleasing from the start of the process and remains so. To achieve this, up to 30 plants per square meter are planted to create a dense green texture, while the green wall, shifted by 90–100 degrees from the conventional landscape architecture project,

Preface

transforms what is usually a carpet into a tapestry. Even the conventional method of introducing plants to a structure – when troughs or pots filled with soil are placed on the floor on the various levels of a building – helped to encourage the current trend to add bushes, grasses or even trees to the façades. This more typical method of greening played an important role in terraced buildings, particularly in the housing estates of the 1970s. While the choice of plants fell to the residents in those days, today's architects take into account in the design stage the image that will be generated by the planting and carefully plan, with the help of gardeners or landscape architects, how the greening will look. Will it be delicate and airy or an impenetrable "Sleeping Beauty hedge"?

Roof greening

Roof greening was also a common element of 1970s architecture and likely featured in the hanging gardens of Babylon too. When it rains, flat roofs can serve as reservoirs, making complicated irrigation procedures – as would have been required in the desert climate of Babylon – unnecessary. In areas which experience high levels of rainfall, roof greening can simplify irrigation. Light is also usually available in plentiful supply, a factor that must be given careful consideration when planning façade greenings. Aside from the high-lying roof greenings, it is the inclined greenings – half façade, half roof, hanging over the whole building – that truly bring a sense of the natural to a building and preserve the image of a landscape. In these examples, there is often a very fluid transition to subterranean parts of the structure. As a remedy for soil sealing, deforestation and a lack of local recreation space, the greening of buildings has become an issue for politicians and urban planners, not just a matter of architecture or esthetics. In the attempt to unify natural elements with the high building density of an urban space, a layer of green is created that rests like a veil over the city.

Interior greening

The third chapter of this book considers interior greenings, which go far beyond the assembly of a few plants or vases of flowers. But like them, the examples featured here are first and foremost good for the soul. Let us bring nature into this space, they cry! Yet they also benefit the interior climate, just as façade and roof greenings improve the climate of the city. The plants purify the air inside the building and absorb formaldehyde, nicotine, benzene and phenol. They also dampen noise and increase humidity, improving the health of the residents, even in heated rooms. Interior greening also demonstrably helps to reduce stress and increase motivation – decisive reasons to introduce greening in offices. In the world of commerce, studies have shown that greened spaces positively influence the consumer experience, increase the time customers spend in the store and boost sales.

Almost all of these artificial habitats, whether interior or exterior, require special provisions in order to thrive, whether structural or through additional watering. These are usually devised by the building or landscape architects and hidden within the architecture itself. This is easiest with roof greenings, where the plants already hang over the sloping walls; here the natural incline is often sufficient to allow any excess rain water to escape and act as a natural sprinkler to water the grass below. It is only soil flushing for which extra provision must be made. The more remote from nature a greening is, whether in the interior or on a façade, the more intensive the preparation must be. Most of the designs featured here include automated irrigation systems. Where possible, along with the scientific names of the plants used, the method of irrigation at the very least is also specified.

This monograph showcases the diverse range of possibilities that exist when plants are used as a building material. Countless different varieties can be integrated in numerous ways, and we see divergent architectural visions realized in an array of contrasting and unique structures. In some cases, the plants seem domesticated, standing neatly in file; in others, they are a symbol of visual creativity; in others still, wildly rampant, their true nature comes to the fore. For building materials, whether plants or something else, can be bearers of meaning, a way of expressing a mood or idea. And even when they have a will of their own, in this case to grow, they can be deployed in a host of different ways as a means of creative expression.

Musée du Quai Branly
Landscape design:
Patrick Blanc. Architect:
Atelier Jean Nouvel.
Paris, France, 2006.

Façades

Is this vertical, façade or wall greening? Is there even a difference between them? Vertical gardens are enjoying great popularity, especially in urban areas as they offer the possibility of creating a green living space. New ways of constructing such gardens appear on the market almost weekly, thrilling us with images from around the world. It is vital to recognize the difference between a garden that is simply optically stunning and one that also utilizes proven technologies. It is also necessary to find out which type of construction can be used in which climate. There are plenty of dramatic images of vertical gardens, but we are only rarely treated to the sight of one in reality. The most important element is the growth of the plants over the course of many years, as only then can the true image of the garden emerge. Even the ecological usage of such a wall reveals itself only gradually. The 90deGREEN system has been demonstrating its viability since 2006 with its patented method, which distinguishes itself from others partly in its capacity to prevent the damaging effects of frost. A specially manufactured mounting plate is used, which, helped by insulation, protects the plant roots from frost. This guarantees the successful long-term development of the plants. 90deGREEN offers a complete greening system, from the wall itself to the plants. This house in Eisenstadt, Austria is just one of many residences boasting a vertical garden designed by 90deGREEN.

Landscape design:
90deGREEN GmbH.
Architecture: Halbritter
& Hillebrand ZT GmbH.
Location: Eisenstadt,
Austria. Year: 2012.
Irrigation: fully auto-
matic, all-year and with
fertilizer.

The Garden for the Wall

Number of species: 14. Plants mainly used:
Santolina, Bergenia, Geranium.

Previous page: the rich
and diverse planting out,
bees on flowers.
This page: plants on the
wall, greenery on the
arch, detail of plants.

Milan's Vertical Forest (also referred to as Bosco Verticale) is the prototype of a radical new architectural concept. The traditional approach to sustainability, based merely on energy efficiency, is replaced with a non-anthropized urban approach, one that aims to increase biodiversity by promoting cohabitation between the human and the natural spheres. Perceived in this context, the Vertical Forest can be understood as a dwelling for trees and birds, inhabited also by humans, a biological habitat that enhances biodiversity. The design promotes the formation of an urban ecosystem where a variety of plant types generates a unique vertical environment, but which functions within the existing biological network, welcoming up to 1,600 species of birds and insects including many butterflies. In this way, the building exists as a catalyst for repopulating the city's flora and fauna. The Vertical Forest also helps to filter fine particles contained in the urban atmosphere, while the diversity of plants creates a microclimate that increases humidity, absorbs carbon dioxide, produces oxygen, and protects against radiation and noise pollution.

Landscape design and architecture: Boeri Studio. Landscape and botanist: Studio Laura Gatti. Location: Milan, Italy. Year: 2014. Irrigation: diverse irrigation methods according to requirements.

Vertical Forest

Number of species: 60. Plants mainly used: Quercus ilex, Jasminum nudiflorum.

Previous page: the Vertical Forest rises above the urban landscape. This page: the lush greenery cascades down the balconies, exterior view, details.

The Green Box

Landscape design:
Gheo Clavarino. Design
and architecture:
act_romegialli.
Location: Cerido, Italy.
Year: 2011. Irrigation:
drop by drop irrigation
on the perimeter of the
ground floor.

The Green Box project was the result of the renovation of a small, disused garage belonging to a weekend house on the slopes of the Rhaetian Alps. A structure made of lightweight metal, galvanized profiles, and steel wires wraps the existing volume and transforms it into a tridimensional support for the climbing vegetation. The greenery comprises mostly deciduous vegetation: Lonicera periclymenum and Polygonum baldshuanicum provide the main texture with Humulus lupulus and Clematis tangutica forming the secondary layer. Groups of herbaceous perennials (Centranthus ruber, Gaura lindheimeri, Rudbekia triloba) alternate with annual varieties (Cosmos bipinnatus, Tagetes tenuifolia, Tropaeolum majus, Zinnia tenuifolia) and bulbous plants to ensure continuous flowering. The interior comprises a room for gardening tools, an area for cooking, and a space for relaxing and socializing. The materials are kept rough and simple: galvanized steel for the kitchen, larch planks for flooring, windows in unpainted galvanized steel, and simple pipes for the water supply. A small green shelter in the vegetation is a privileged space for observing the changing seasons in the surrounding park. Left wild in some areas and transformed into a garden of flowers or a simple green space scattered with rocks in others, the park is a joy to behold.

Number of species: 11. Plants mainly used: Cosmos bipinnatus, Tagetes tenuifolia.

Previous page: soft light falls into the kitchen through the lush greenery, the structure is disguised by vegetation. This page: the house blends perfectly into the surrounding park, glass sliding doors allow for sunlight to flood the rooms, building plan.

24
The Rest

Landscape design: Leon Kluge Garden Design. Architecture: GJ Architects. Location: Nelspruit, South Africa. Year: 2015. Irrigation: automatic.

This garden reflects the design and color scheme both inside and outside the house. Bright red is the prevalent color in the kitchen overlooking the vertical garden. The pool area next to the garden is also red in color. The vertical garden softens the building but also serves as a piece of art that can be viewed from most parts of the house.

The red spirals in the design echo the color of the interior of the house, while their dynamic shape contrasts with the square building, enhancing their esthetic potency. The metal spirals provide the color for the vertical garden, as the plants used are mainly shades of green.

Previous page: view on
the house. This page:
overlooking the garden,
view on the vertical
garden, view from the
driveway, details of the
garden.

Number of species: 9. Plants mainly used: Bromeliaceae, Tillandsia.

Green Renovation

Landscape design and architecture: Vo Trong Nghia Architects.
Principal architects: Vo Trong Nghia, Takashi Niwa, Tran Thi Hang.
Contractor: Wind and Water House JSC.
Location: Hanoi, Vietnam. Year: 2013.
Irrigation: automatic mechanical irrigation.

Greenery and abundant light revitalized this dilapidated house. The massive concrete staircase was replaced with slender steel stairs, creating space for a triangle light well made of monumental marble, through which natural light can fall from the outside. A second void was cut into the first floor, connecting the dining space with a study, inviting communication between the residents and bringing sunlight into the house. The ground floor has been raised to install a layer of air ventilation beneath and to prevent condensation. A green façade called Greenfall, reminiscent of a waterfall, is the defining feature of the residence.

The old fences on the balcony were removed and replaced with galvanized steel trellis. Residents and guests can enjoy views of the greenery from every room in the interior. The green façade and the roof garden function together to reduce energy consumption and to protect the building from harsh sunlight. Many kinds of vegetables, flowers and even a tree have been planted on the roof. This greenery system is prototypical and applicable to all buildings in tropical climates. The designers created this residence as a model for greening tropical cities where the benefits of a healthy green home can be enjoyed by the occupant and have a wider positive impact on the city.

Number of species: 10. Plants mainly used: Combretum indicum.

Previous page: view on the Greenfall, bedroom. This page: the greened residence is a refreshing addition to a street of more conventional dwellings, the rooftop is a green oasis, even in the wood-clad living area residents can enjoy views of the greenery.

The Algae House

Landscape design and architecture: SPLITTERWERK. Building services engineering: Arup, Immosolar. Structural engineering: Bollinger and Grohman. Location: Hamburg, Germany. Year: 2013.

Born out of the International Architecture Exhibition in Hamburg in 2012 and 2013, the groundbreaking case study house features the world's first algae bioreactor façade. This smart material house combines intelligent materials and technologies with new typologies of living. Both algae bioreactor façades face the sun and feature bold red and white stripes, recalling the colors of the Hanseatic city. But above all, these stripes of color highlight the singularity of this apartment block and its role as a radical prototype for photobioreaction as a generator of energy and a means of controlling light and shade. While the continual growth of algae generates an impression of ever-changing color when viewed from a distance, up close the façades themselves seem to move and shift. The bubbling motion created by the supply of carbon dioxide and nitrogen and the circulation of water, a constant requirement for the suspended algae plants, invite one to view this revolutionary production of biomass as a solar-generated art installation, quietly bubbling away to itself. Legible from a distance, the question "Photosynthese?" ("Photosynthesis?") and the response "Cool!" are plastered in black writing across the two green façades with tiny windows on the shaded side of the structure. The final flourish is provided by the ornamental stucco grapevines entwined across the penthouse façades at the top of the building.

Number of species: 1.
Plants mainly used: microalgae.

Previous page: solar
generated art installation
or the manufacture of
biomass. This page: the
red and white stripes are
a bold addition to this
already radical structure,
functional diagram.

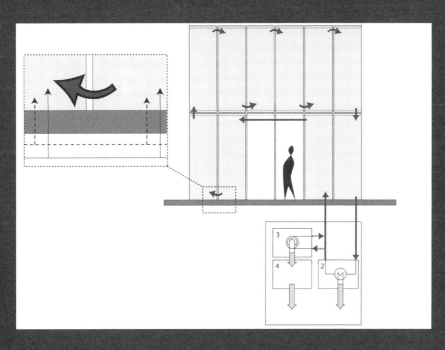

36

One Central Park

Landscape design:
Patrick Blanc,
Jean-Claude Hardy,
Aspect Oculus. Archi-
tectural design: Ateliers
Jean Nouvel. Local collab-
orating architect: PTW
Architects. Location:
Sydney, Australia.
Year: 2015. Irrigation:
drip irrigation.

The building's key features are its hanging gardens, a cantilevered heliostat, an internal water recycling plant and a low-carbon tri-generation power plant. The landscape artist designed the 1,120-square-meter vertical garden that covers the surface of the building. The suspended oasis boasts 35,200 plants of 383 different species, including some natives such as acacias. The gardens use a remote controlled dripper irrigation system and a special process developed by Blanc, whereby plant roots are at-tached to a mesh-covered felt soaked with mineralized water. This allows the plants to grow without soil along the façade. The gardens are maintained by the local green roof and wall company Junglefy. The company was forced to replant some of the building's gardens in 2015 after a water source was accidentally cut and the system's alarm failed. Gray water is currently piped into the laundry and bathroom areas within the apartments, and used to water the external green spaces.

Number of species: 383. Plants mainly used:
Acacia plicata.

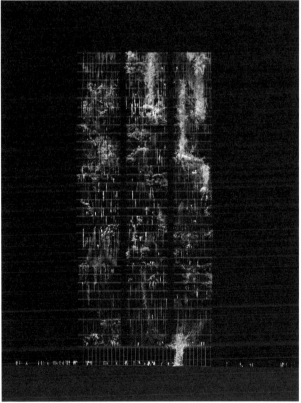

Previous page: exterior
view. This page: view
from the sky garden
towards city, planted
terrace, eastern façade.

IAC 8800 Sunset Boulevard

Landscape design and
architecture: Rios
Clementi Hale Studios.
Consultant: Rana Creek
Design. Location: West
Hollywood, USA.
Year: 2012.

IAC, a leading media and Internet company, selected Rios Clementi Hale Studios to upgrade the exterior of its West Hollywood offices. The architects' collaboration with Rana Creek saw the creation of a living canopy of plants that extends up the street façade of the seven-story brick building. The green wall consists of a five-story trellis landscaped with more than 11,000 plants, which infuses a busy commercial intersection with a connection to the surrounding hills and supports the native habitat by attracting hummingbirds, bees, and butterflies. An ingenious irrigation system economizes 378,542 liters of potable municipal water per year by re-using gray runoff that was previously daily pumped into sewage from the building's garage. The project includes publicly accessible landscaped plaza and a new entry lobby to correspond with the enlivened building. The steel latticework frame projects five meters in front of the façade to form a dramatic canopy over the building entrance and a café on the site. This tapering structure, bolted to the concrete structural frame, reaches a height of 22 meters and widens to about 46 meters at its base. Its open grid, designed to correspond with the building module, preserves views while covering the 1980s brick façade, updated with a simple coat of white, forging an ever-changing display of nature. LED lighting fixtures embedded in the lattice illuminate the planted wall at night, emitting a lantern-like glow.

Number of species: 29. Plants mainly used: Epilobium canum, Fragaria chiloensis.

Previous page: greened
façade, general view.
This page: exterior, view
down from window, view
from inside, site plan.

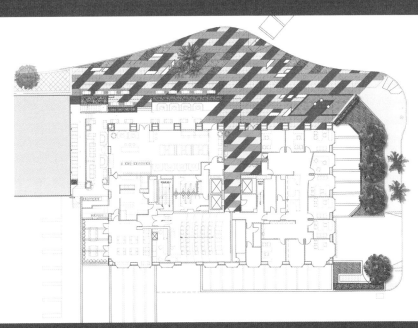

44

Green Cast

Landscape design and architecture: Kengo Kuma & Associates. Structural engineering: Makino Structural Design. Location: Odawara-shi, Japan. Year: 2011. Irrigation: automatic watering system.

Kengo Kuma & Associates covered the façade of the new mixed-use building in Odawara in aluminum die-cast panels, which frame the views and shield the functional spaces behind an otherwise transparent front. The monoblock cast panels are placed in groups of three to six and have a double function as planters. Each panel is slightly slanted, and the organic appearance of the surface is derived from decayed styrene foam. Equipment such as watering hoses, air reservoirs for ventilation and downpipes are installed behind the panels, allowing the façade to accommodate a comprehensive service system. Principal tenants include retail, a clinic, offices and private residences. This way, the greenery benefits a wide range of people, whose daily routines bring them in the proximity of the eye-catching, lively façade.

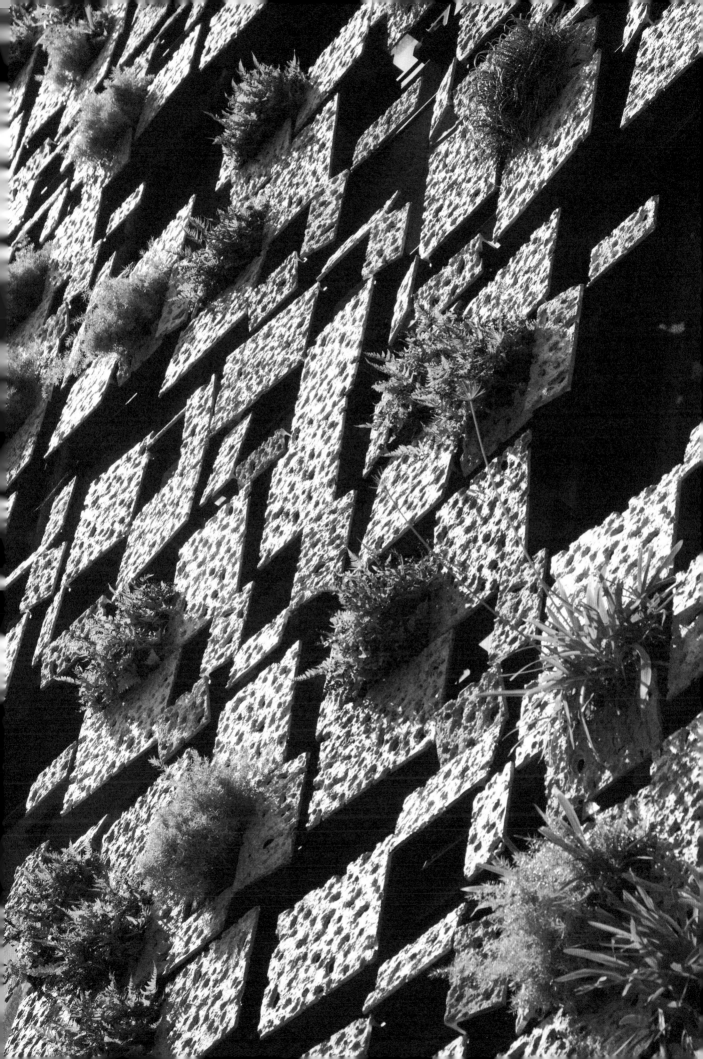

Number of species: 5. Plants mainly used:
Asparagus officinalis, Ophiopogon japonicus.

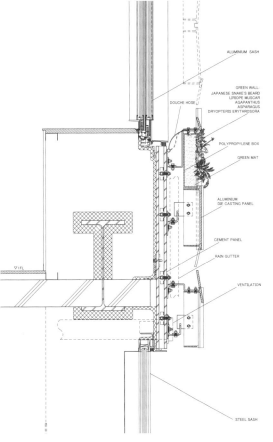

ALUMINIUM SASH

GREEN WALL
JAPANESE SNAKE'S BEARD
LIROPE MUSCAR
AGAPANTHUS
ASPARAGUS
DRYOPTERIS ERYTHROSORA

DOUCHE HOSE

POLYPROPYLENE BOX

GREEN MAT

ALUMINIUM
DIE CASTING PANEL

CEMENT PANEL

RAIN GUTTER

VENTILATION

IFL

STEEL SASH

Previous page: close-up
of the façade. This page:
view of the staircase,
general view, plan.

With an area of over 670 square meters, space for 150 seated guests and room for 100 more in the outdoor dining area, the glass building known as Green Cube is a major addition to the town of Bielefeld. The catering areas are spread over the ground and upper floors, while half of the flat roof area is designed for use as a roof terrace. The core structure consists of an almost completely glazed restaurant area and a business zone with a brick façade. Like a second skin, a green layer of carefully selected evergreen ivy plants envelops the core in a living, breathing embrace. On the three sides facing the square, certain elements protrude from the green façade. The south side features the main entrance, while the upper floor on the eastern side boasts a balcony that opens the building towards the square and protects the outdoor dining area from sun and rain. A bay window on the north side offers views of the fountain, a popular feature of the site. In the interior, an expansive open space connects and enables views between all three levels. The first floor is configured for shorter visits and has the character of a bistro. Stone was chosen for the floor of this intensively used area, while the upper level features oak floorboards and has the atmosphere of a restaurant. The possibility of dividing the area into two separate spaces introduces a welcome degree of flexibility.

Landscape design and
architecture: Detering
Architekten. Further
participants: BGW,
Laskowski Ingenieur-
gesellschaft mbh,
zwischen_raum,
Ingenieurbüro Deymer.
Location: Bielefeld,
Germany. Year: 2014.
Irrigation: automatic.

Green Cube

50

Number of species: 1. Plants mainly used: Hedera helix.

Previous page: view
on the green cube, the
bar as the heart of the
building. This page: view
on the ivy-clad terrace,
exterior views.

52
CSI-IDEA Building

Landscape design: Paisajismo Urbano. Architecture: Juan Blázquez+Oficina Técnica Municipal. Location: Málaga, Spain. Year: 2015. Irrigation: computerized and sectored system.

The design of this building was conceived as a prototype of the Zero Energy Buildings concept, which will be mandatory for all new buildings from 2020 onwards. The building features a host of sustainable design features including passive design elements to reduce energy demand, efficient lighting and air conditioning, the use of solar thermal energy for heating, and the integration of vegetation as a construction element. All of the necessary power is generated within the building, making it a plus-energy building, while many of the materials are recycled and recyclable, and water consumption and waste generation are kept to a minimum. This truly is an ecologically sustainable building. Consideration was also given to the impact of the building's construction on the neighborhood. It has been elevated to allow the existing park to pass under the building and vertical gardens have been planted on the façades facing the nearby homes.

54

Number of species: 25. Plants mainly used:
Salvia divinorum, Tulbaghia.

Previous page: view on
the lush green exterior
wall. This page: the
spectacular architecture
as a true eye-catcher,
details of the greenery,
view from a side angle.

Tuen Mun Hospital has been serving the community for almost 30 years. SK Yee Healthy Life Centre aims to provide a modest yet sustainable space within the existing hospital complex providing high-quality counseling to patients. This new center is far more than a healing environment; envisaged as a home, a garden and a playground, the space promotes healthy living for all who encounter it, whether patients, healthcare professionals or local residents. Ronald Lu & Partners (RLP) adopted a lean-and-green design that has revolutionized the existing rooftop and imbued it with new meaning. The rooftop center offers an ambience of calm and serenity throughout, immersing patients in nature and daylight and offering them a stress-free experience. Each counseling room and functional area is adjoined to a garden, creating a constant interplay between interior and exterior spaces. Natural greenery, light and air are welcomed into this health-giving structure. The design transforms the roof into a hub comprising a reception lounge, four counseling rooms, a multipurpose room and a roof garden comprising over 57 percent greenery, offering a landscape for the enjoyment of both the patients and the center's neighbors. The extensive green roof also provides welcome thermal protection to the ward below.

Landscape design and architecture: Ronald Lu & Partners (RLP). Location: Hong Kong, China. Year: 2014. Irrigation: auto-irrigation system.

SK Yee Healthy Life Centre

Previous page: lush greenery dominates the exterior from the walls to the green roof.
This page: an empty roof is transformed into a meaningful space, the interiors are flooded with air and light, the children's counseling room is designed with their particular needs in mind.

Number of species: 3. Plants mainly used: Plumeria rubra, Zoysia tenuifolia.

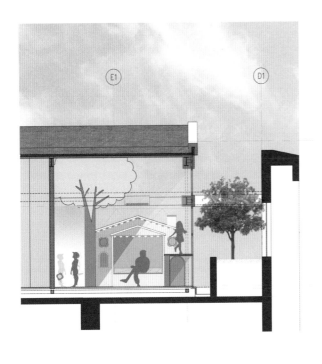

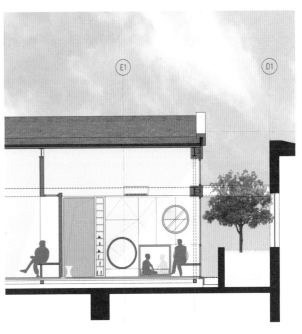

The Center for Early Education

Landscape design and
architecture: Rios
Clementi Hale Studios.
Irrigation design:
Sweeney & Associates.
Structural design:
Englekirk Structural
Engineering. Location:
West Hollywood, USA.
Year: 2012.

The architects and client worked together to apply landscape programming to the challenge of creating an efficient, engaging, and educational solution that brings a series of ecological learning moments to young children. The resulting features – or interventions – involve the school's main entryway, circulation paths, rooftop, and underground garage. The interventions include a distinctive, planted entranceway that forms a vertical landscape, or living wall; pockets filled with native plants hanging from main outdoor circulation stairway railings; a rooftop turret containing a solar-powered weather station and an interactive wind turbine; as well as an irrigation system that is fed from a collection tank in the underground garage to irrigate plants throughout the campus, while demonstrating water use and conservation.

The design connects the school to its local and regional environment by referencing plant ecologies from the surrounding areas that are visible from the school's rooftops, as well as by celebrating the phenomenon of the bountiful watershed flowing underneath the school. The experience begins with the vertical garden surrounding the school's main entrance – a "green threshold" formed out of 34 vertical planter boxes that contain a profusion of California succulents.

Number of species: 38. Plants mainly used: Rhamnus californica, Ramirez Begonia, Polystichum munitum.

Previous page: detail of the plants, entrance area. This page: title on greenery, staircase, detailed view, project explanation.

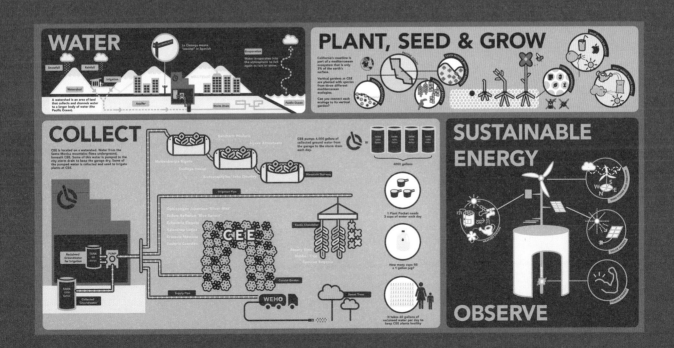

Grabouw

Landscape design: Leon Kluge Garden Design. Architecture: Alan Walt architect | associates. Location: Grabouw, South Africa. Year: 2011. Irrigation: automatic.

Located in the hills of the Cape Peninsula, this residence experiences a climate of high winds and dry summers. A host of plants rescued from the building site before bulldozing began were lovingly integrated into the design. The garden was designed to be a piece of art, a tapestry of color and texture. A flock of birds chose to nest in the vertical wall, bringing new life to the already lush façade. The interior boasts a modern esthetic, its bright colors echoing the vibrant color palette of the vertical garden. It took the designers a month to erect and complete the wall. Agapanthus africanus (African lily), Juncus effusus (soft rush) and Dietes bicolor (African iris) are just a few of the featured plant species.

Previous page: the water used to maintain the vertical garden is recycled and reused. This page: Grabouw proves that greening can be artistic.

Number of species: 20. Plants mainly used: Agapanthus africana, Juncus effusus.

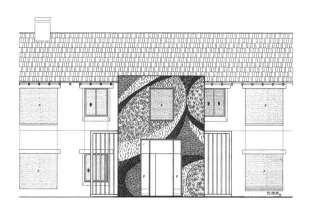

In December of 2014, Antonio Maciá and Paisajismo Urbano completed the construction of the Calahorra Gastrobar. This cafeteria is composed of a 120-square-meter terrace and a small 20-square-meter building, covered entirely by a vertical garden. This combination of vertical garden and cafeteria is located next to the party wall of the structure, which adjoins La Calahorra, one of the city's historical buildings. Lacking a defined use until the onset of this project, this small urban corner of the city has been revitalized. It is now a reference point for locals and a tourist attraction for visitors who are drawn by the vibrant and colorful living wall. More than 3,000 plants, belonging to 15 species from five different families, cover an area of 100 square meters. Most of them are native to the region. With La Calahorra, Paisajismo Urbano managed to create a modern urban habitat that is populated by humans, plants and animals alike.

Landscape design:
Paisajismo Urbano.
Architecture: Antonio
Maciá Arquitectura y
Diseño. Location: Elche,
Spain. Year: 2014.
Irrigation: computerized
and sectored (NPKSYS-
TEM-5000).

La Calahorra

Number of species: 15. Plants mainly used:
Rosmarinus officinalis, Lavandula angustifolia.

Previous page: the
living wall rises high,
plantation plan.
This page: the vertical
garden after two
months, detail of plants.

Roofs

Seongbuk-dong Residences

Landscape design
and architecture:
Joel Sanders Architect.
Executive architect:
Haeahn. Location: Seoul,
South Korea. Year:
2010. Irrigation: dripper
irrigation system.

Located in the exclusive Seongbukdong district of Seoul, this enclave of 12 sustainable houses is designed to take advantage of its steeply sloping site by updating the ancient principle of the "borrowed view". The staggered arrangement of L-shaped dwellings ensures that each unit enjoys unobstructed open southern views of a wooded valley, framed in the foreground by its neighbors' planted green roof. In addition, each house possesses apertures that frame views of its neighbor's rear and side yards, visually expanding the property of each unit without compromising privacy.

Planted with a multicolored pattern, the roof of each house merges to form a dramatic composition that changes color with the seasons and is visible from the public park across the valley. Each of the four unit types is organized around two terraces that spatially and visually connect the interior with the exterior. On the main floor, cutouts in the floating terrazzo floor channel greenery from the lower terrace to a dramatic top-lit interior stairwell that links the living and dining areas with a second terrace shared by the bedrooms above.

Number of species: 18. Plants mainly used: Prunus mume.

Previous page: the steeply sloping hill faces a public park. This page: the ancient principle of the "borrowed view" is replicated here, the large living room, a street façade reminiscent of Seoul's traditional urban fabric, view on greened roofs.

The California Academy of Sciences in San Francisco is one of the few natural sciences institutes where public experience is directly related to in-house scientific research, which has been performed in the original building ever since its foundation. The new Academy is designed on the site of the previous facility in Golden Gate Park. This project required the demolition of most of the 11 existing buildings erected between 1916 and 1976. The mission statement of the Academy, ("To Explore, Explain and Protect the Natural World"), together with San Francisco's mild climate, made this project ideal for incorporating sustainable design strategies. In addition to energy efficient heating and cooling, a more holistic approach was agreed upon, involving exerting a serious effort in choosing building components, recycling of demolished structures' materials and pioneering construction processes with a view of lifecycle efficiency.

The green roof is landscaped with native plant species which are drought resistant and do not require irrigation once established. The roof extends beyond the perimeter walls to become a glass canopy providing shade, protection from rain, all the while generating energy using photo voltaic cells embedded in the glass. At the center of the living roof is a glazed skylight covering a piazza. Smaller skylights scattered across its surface allow natural light into the exhibit space and automatically open for natural ventilation.

The positioning of spaces in relation to daylight and ventilation is integral to the building design, as are water and run-off efficiency and energy generation.

Landscape design: Rana
Creek Design. Archi-
tecture: Renzo Piano
Building Workshop with
Stantec Architecture.
Location: San Francisco,
USA. Year: 2008.

California Academy of Sciences

Previous page: overview
of the entrance area.
This page: general view,
bird's-eye view, plans,
focus on the rooftop.

Number of species: 30. Plants mainly used: Fragaria chiloensis, Prunella.

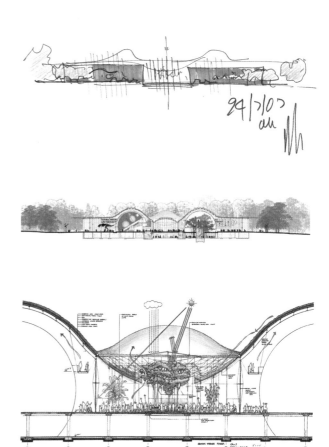

Villa Topoject is located in a small valley in a mountainous area near Seoul, Korea. The valley slopes upwards, scattered intermittently with rows of houses facing a single lane road. The designers chose to forego walls and fences on this side, instead using the topography itself to generate private interior spaces. The building opens up towards the southern side and towards a small creek. Villa Topoject is a home for a couple who want to enjoy rural life and preserve their relationship with the land while still being able to commute easily to the city. The residence is small but full of rich details. The design represents a rejection of the dichotomy of object building versus landscape building. A hybrid of the two, it comes alive in the fusion of two architectural typologies. The topography itself becomes an object, creating semi-private outdoor spaces, while the interior offers places of perfect privacy. The continuous exterior spaces meet the interior at all levels adding rich spatial qualities. The boundary between exterior and interior, land and building, subject and object is rendered ambiguous in this dramatic structure.

Landscape design and
architecture:
Architecture of Novel
Differentiation (AND).
Location: Yangpyeong-
gun, Gyeonggi-do, South
Korea. Year: 2010.
Irrigation: sprinklers.

Villa Topoject

Previous page: the angular architecture makes the building stand out from the soft and green nature, aerial shot of the spectacular building. This page: view on the swimming pool, first and second floor plan, the backyard is for the slow life of the residents, a magnificent indoor garden brings nature to the inside of the house as well.

Number of species: 20–30. Plants mainly used: Betula populifolia, Hydrangea.

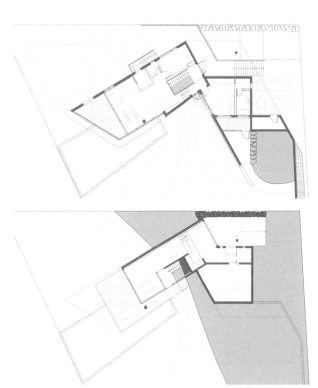

Via Verde – The Green Way

Landscape design and architecture: Dattner Architects, Grimshaw Architects. Location: New York City, USA. Year: 2012. Irrigation: rain water. Further facts: affordable housing project.

Via Verde – The Green Way stands as a symbol of what is possible for future developments in terms of sustainable practices and healthy living in urban environments. The building was conceived as a means of boosting mixed-use and mixed-income residential development, which is achieved by providing a pharmacy and community health clinic as well as live/work units at street level. The building features a total of 222 residential units, 71 of which are for middle-income households. The remaining units are evenly spread between low- and moderate-income rentals. The whole complex consists of a 20-story tower and a mid-rise building with duplex apartments and townhouses. Several roof terraces, spiraling down from top to bottom, form the pivotal design concept. Rainwater is collected and stored on the premises for reuse. Sustainability and eco-friendliness were the architects' central design principles.

Number of species: 49. Plants mainly used: Solanum lycopersicum, Raphanus sativus.

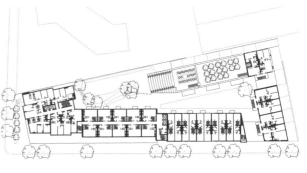

Previous page: the
complex is open and
modern. This page: the
city skyline forms the
backdrop to the building,
site plan, residents help
maintain the green spaces.

Edgeland House is located on a rehabilitated brownfield site and is a modern re-interpretation of one of the oldest housing typologies in North America, the Native American pit house. The pit house, typically sunken, takes advantage of the earth's mass to maintain thermal comfort throughout the year. Like this timeless dwelling, Edgeland House's relationship to the landscape both in terms of approach as well as building performance involves an insulating green roof and a seven-foot excavation, gaining benefits from the earth's mass to help it stay cooler in the summer and warmer in the winter. Such an architectural setting presents an opportunity for maximum energy efficiency. Edgeland House is about healing the land and ameliorating the scars of the site's industrial past. The project raises awareness about a diminishing natural landscape and its finite resources by creating a balance between the surrounding industrial zone and the natural river residing on the opposite side of the site. Both visually and functionally, Edgeland House touches on architecture as site-specific installation art and as an extension of the landscape. The program is broken up into two separate pavilions, for the living and sleeping quarters, and requires direct contact with the outside elements to pass from one to the other. This project sets new standards for sustainability while providing sublime esthetic qualities through its small footprint and integrated mechanical features. Lady Bird Johnson Wildflower Center collaborated to reintroduce over 40 native species of plants and wildflowers to the Edgeland House green roof and site, serving to help protect the local ecosystem.

Landscape design:
Ecosystem Design
Group Lady Bird Johnson
Wildflower Center.
Architecture: Bercy
Chen Studio.
Further participants:
MJ Structures, Water-
street Engineering.
Location: Austin, TX,
USA. Year: 2012.
Irrigation: drip irrigation.

Edgeland House

Number of species: more than 40. Plants mainly used: Lupinus texensis, Schizachyrium scoparium.

Previous page: view on
the terrace, entrance.
This page: view from the
roof, terrace, view on the
entrance.

The Corda Campus in the Belgian city of Hasselt is both a technology center and a modern workplace where young and innovative companies can thrive. At the heart of the campus is the service center Corda 1, containing offices, conference rooms, a restaurant, café and many other services. With its striking greened steeply pitched roof and the combination of concrete,

glass and green, the Corda 1 building is also the site's visual centerpiece. Both of the site's greened pitched roofs integrate modern architecture and design with natural elements and form the core elements of the campus' "green lungs". The two greened buildings with their combined pitched roof area of 2,300 square meters have a slope of 20° and a maximum flow

length of up to 70 meters. With this vast surface area in mind, the client and architect decided to cover the roof with a partly accessible grass lawn. Various challenges soon followed, involving both the construction of the feature itself and the practicalities of introducing vegetation. These could only be overcome by utilizing a tried and tested system of greening. The designers chose the Optigrün system "Pitched roof Type P" featuring the Optigrün Drainage Board FKD 58 SD. Functions of this system also include shear stress distribution and water storage.

Landscape design:
Optigrün international
AG, Ecoworks Vilvoorde.
Architecture: ELD.
Further participants:
Democo. Location:
Hasselt, Belgium.
Year: 2014. Irrigation:
automatic drip tubing.

Corda Campus Hasselt

Number of species: 1. Plants mainly used: grass.

Previous page: Corda
Campus Hasselt, view on
the greened roof area.
This page: looking down
from the highest point,
side view, grass fields.

The new magnetic resonance imaging (MRI) laboratory is part of the "physical park" at the National Metrology Institute in Berlin-Charlottenburg. The MRI unit is installed in the laboratory which was previously used for experiments with the first German quartz clock and which is close to the basement floor of the historical observatory. A green roof un-folds, merging with the park, forming the sculptural entrance to the laboratory itself and revealing its subterranean location. Visitors, test persons and employees enter the new structure via a lobby that faces the observatory. All rooms are located on the lower level but are filled with natural light and air thanks to a sloping inner courtyard. From the interior, views of the observatory and the park's many trees can be enjoyed. The glass façade features molding flush to the surface and the pillars are clad in polished stainless steel. Reflective and insubstantial, almost chimerical, these elements contrast with and complement the concrete roof construction.

Landscape design:
Stefan Bernhard
Landschaftsarchitekten
with Philipp Sattler.
Architecture: huber
staudt architekten bda.
Location: Berlin, Germa-
ny. Year: 2012. Irrigation:
pop-up irrigation.

National Metrology Institute

Number of species: 1. Plants mainly used: turf.

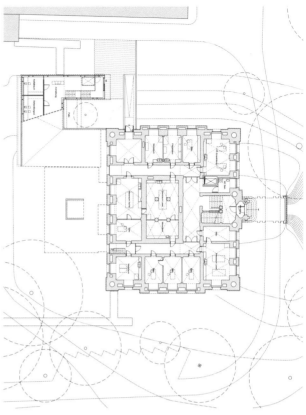

Previous page: observatory with new MRI laboratory, atrium and entrance area.
This page: view from west, ground floor plan, entrance MRI laboratory.

Fusing architecture and nature, culture and history, the Moesgaard Museum offers a holistic design experience. With its green roof, sunlit courtyard gardens and underground terraces, the museum is uniquely positioned to welcome alternative kinds of exhibition. The 16,000-square-meter museum is located in the hilly landscape of Skåde, south of Aarhus. With its sloping roofscape of grass the building appears as a compelling visual landmark, visible even from the sea. Seeming to almost grow out of the landscape, the rectangular roof plane forms an area for picnics, barbecues, lectures and traditional Midsummer Day bonfires during the warmer months. Sustainability is integrated into every aspect of the design. Fundamental elements such as the building's geometry and orientation were considered in order to maximize the efficiency of every square meter. The south-facing roof surface – or roof façade – was conceived as the central element for this highly energy efficient building, which has been awarded Energy Class 1 status. The green roof contributes to the building's efficient energy consumption, reducing the overall need for cooling due to increased heat absorption. It also minimizes the amount of wastewater draining from the site. Sloping downwards towards the south, the roof protects the objects on display from direct sunlight. Adjoined to each exhibition room, a glass-enclosed area functions as a lobby, allowing visitors to enter, but preventing sunlight from reaching the exhibited items. In these spaces, visitors can enjoy a respite from the dark of the exhibition rooms and reorient themselves to nature and sunlight.

Landscape design:
Kristine Jensens
Tegnestue. Architec-
ture: Henning Larsen
Architects. Further par-
ticipants: , Cowi, D-K2,
MT Højgaard og Lindpro.
Location: Aarhus, Den-
mark. Year: 2010-2013.
Irrigation: none.

Moesgaard Museum

Previous page: the museum's lush green sloping rooftop, the structure merges seamlessly with its surroundings.
This page: breathtaking views of the countryside and the azure sea, ground floor plan, the entrance area is grand yet inviting, cross section.

Number of species: 1. Plants mainly used: grass.

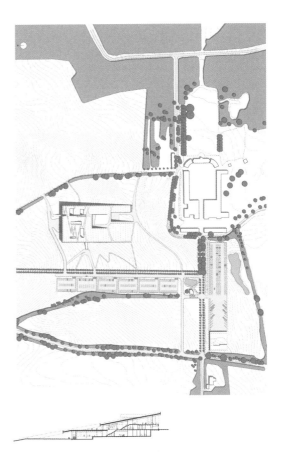

The coast of Katwijk has been designated a weak link in the Netherlands' battle against the sea's encroachment. To keep Katwijk and the hinterland safe for the next 50 years, coastal reinforcements are necessary. In collaboration with Arcadis, Katwijk and the Water Board of Rhineland, OKRA has designed this innovative form of coastal reinforcement. Inspired by the scenic village atmosphere, the designers preserved the character of the boulevard and applied it to the new shoreline. The design generates a harmonious relationship between the sea, the family resort and accommodation within the new coastal zone. The dune transitions to the beach are positioned in line with the street pattern of the village, ensuring visual continuity and complementarity. Following a curving pattern, the network of longitudinal and transverse paths echoes and enhances the esthetic quality of the dunes themselves. Lookout points and a dune boardwalk offer a view of the sea where this is lost with the broadening of the coast. The hidden underground parking area has a floor area of 14,000 square meters and comprises 11,500 square meters of greened surface. This recreation area was covered with over 57,500 Marram Grass plants.

Landscape design: OKRA Landschapsarchitecten. Architect underground parking: Royal HaskoningDHV. Client: Municipality of Katwijk, the Water Board of Rhineland. Location: Katwijk, the Netherlands. Year: 2015. Irrigation: none.

Kustzone Katwijk

Previous page: the new structure peers out across the landscape, various geometric forms work in harmony in the design of the square and entrance way.
This page: the access points welcome visitors to this architecturally innovative space, plans.

Number of species: 1. Plants mainly used: Ammophila.

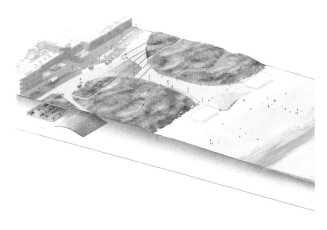

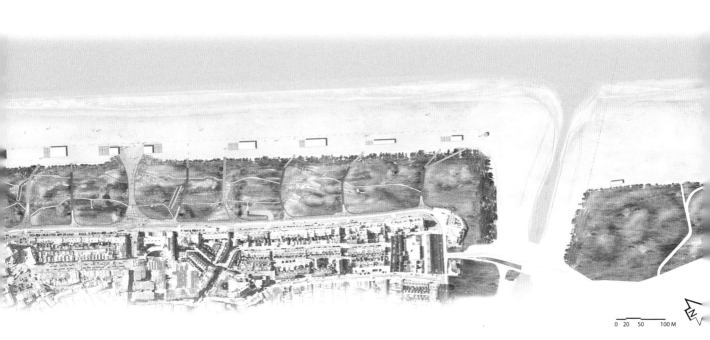

0 20 50 100 M

The Marcel Sembat High School offers technical education about motors and vehicle mechanics. It demands vast spaces with high ceilings and large surfaces. It's located in a declining industrial suburb of the port of Rouen. The site and the size of the program turned the project to a combination of landscape design, architecture and town planning. While cleared of the six early disordered 1930s buildings, the site appears like a clearing between a large park and the city. The needed substantial volume for the workshops is covered by a striped green roof. The building starts at the boundary of the park and fits naturally in the site by the wavy design of its vegetated roof. It reconnects the existing school with its surroundings, so that its soft lines and slopes blend naturally into the physical features of the park on one hectare. The architects aimed to use the largest and lightest part of the site to house the workshops. The classrooms are displayed along the long side of the workshop, with direct visual relations, to bring back together intellectual and manual activities. The design of the roof lets natural light reach deep inside and protects from the direct sun. The curves of the structure, reminiscent of the shapes of vintage car wings, offer a new image of manufacturing environments. The steel structure allows long span use to free the ground level of the structure. On the city side, the architects redesigned the public space to connect the high school with its workshops and with the future public library. They dare that this clear and understandable design will light up students' and teachers' everyday life.

Landscape design and
architecture: archi5.
Further participants:
B. Huidobro, IOSIS
Ouest, SPEC, ABC
Décibel, Constructions
Métalliques Charondière,
Millery Colas. Location:
Sotteville-lès-Rouen,
France. Year: 2011.

Marcel Sembat High School

Number of species: 2. Plants mainly used:
Sedum, grasses.

Previous page:
high-angle shot of the
building, side view. This
page: view from the roof
towards the park, detail
of grasses, view over
the roofs.

The Science City is the result of a dynamic reconstruction of the Hönggerberg campus of the Swiss Federal Institute of Technology in Zurich. Part of the reconstruction comprises new sports grounds on the eastern edge of the site close to Käferberg, a wooded mountain overlooking the city. Designed in harmony with the topography, the building's structure is integrated into the hillside with only its western façade visible. Seeming to flow through the expansive foyer and up a greened ramp, exterior mingles seamlessly with interior. Attractive interior configurations generate a pleasant ambience for teaching, sports training, leisure activities, relaxation and even for big events and celebrations. The smooth structure with its green crystalline hue houses a vast gymnasium with a stage and backstage areas as well as halls for dance and gymnastics, weights rooms and a recovery area. Functional areas are positioned around the halls on all sides and on multiple floors. A partly opaque, partly transparent façade made of green thermally insulated glazing evokes the impression of a crystalline body, smooth in some places, rough in others. The interior configuration allows the flow of people to be absorbed and individuals redirected towards the various functional areas, while the green room and green ramp entice the visitor to re-engage once more with the beckoning exterior world.

Landscape design and architecture: Dietrich | Untertrifaller | Stäheli Architekten. Location: Zurich, Switzerland. Year: 2009.

ETH Sport Center

Number of species: 1. Plants mainly used: grass.

Previous page: bird's-eye
view of the building,
interior view. This page:
corridor facing the
greened slope, exterior
view, staircase facing the
greened slope.

This low-cost, self-assembled residence has been constructed in León for a family of winemakers. At under 400 euros per square meter, the costs were less than half the usual rate for such a house. The design revolves around a series of pavilions, interconnected in a manner inspired by the traditional Chinese architecture of Suzhou Gardens. The structure consists of load-bearing walls made out of lightened clay that provide the house with thermal insulation. The beam system – made of precast concrete and clay – comprises five-meter spans and supports the habitable green roof. Thanks to its soil and vegetation, the roof not only provides insulation but also absorbs solar radiation. During the winter, heating is provided by low-temperature radiant floor heating systems connected to solar panels. In summer, the interior is cooled by cross ventilation at night and by blinds during the day, a feature typical of traditional Spanish architecture. The structure's varying levels generate a series of diverse but interconnected spaces. The diagonal alignment of the pavilions and axial continuity of doors ensure a sense of flow between these spaces. Throughout the design and construction process, the clients were fully engaged in redefining the house's geometry, adapting it to their needs and gradually colonizing the space. Low-cost, simple, traditional construction and building techniques, familiar to locals and the home's owners, were employed throughout the process.

Landscape design and
architecture: Alarcón
+ Asociados / Alberto
Alarcón. Further partici-
pants: Sara Rojo, Carlos
Tomás, Clara García,
Heloise (Architects)
Location: León, Spain.
Year: 2009. Irrigation:
precision nozzles.

Chinese House at León

Previous page: spectacular views of the surrounding landscape. This page: geometric forms offset and complement the natural surroundings, the Chinese House rests resplendent in the warm sunshine, ground floor plan, cross sections.

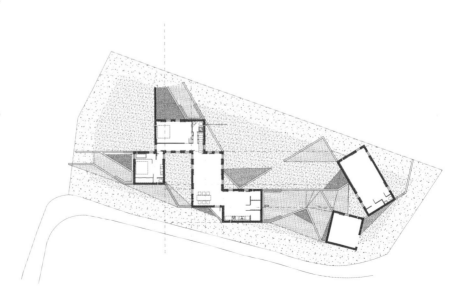

Number of species: not to be determined.
Plants mainly used: regional plants.

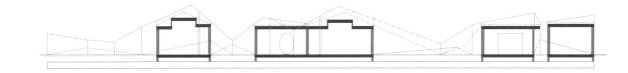

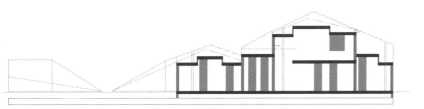

StreetDome is a vast and unique urban landscape for activity and recreation that boasts a 4,500-square-meter skate park and facilities for parkour, boulder climbing and canoe polo. The project's ambition was to set new standards for the design of urban arenas for unorganized sports. It is intended as an open playground and social meeting place for people of all ages,

skill levels and cultures. The StreetDome itself is based on CEBRA's "igloo hall" concept: a low-cost sports hall. To reduce running costs, the hall is unheated and lit primarily by natural light sources, while the building's surface area has been minimized through its dome shape. The roof spans around 40 meters and allows for a large open space free of load-bearing

structures, which can be fitted out for a wide range of activities. The StreetDome's roof is covered with pre-fabricated moss sedum elements, which consist of a variety of stonecrops and spontaneously occurring moss. The green roof serves both esthetic and functional purposes. Visually, the roof changes color between green and a rusty red according to the season and weather conditions, while the appearance of the vegetation will develop over time, influenced by the local climate and fertilization. From June to September the roof blossoms with yellow, white and pink flowers. Functionally, the green roof contributes to an improved climate. As the hall is uninsulated, the vegetation prevents overheating during the summer, acts as sound insulation, filters the air and reduces particle pollution. In addition, the roof is able to absorb rainwater, which then evaporates instead of running off the building and onto the skate park.

Landscape design
and architecture:
CEBRA architecture.
Further participants:
Glifberg+Lykke, Rambøll,
Hoffmann, Grindline.
Location: Haderslev,
Denmark. Year: 2014.
Irrigation: rain.

Haderslev
StreetDome

Number of species: 12. Plants mainly used:
Sedum acre, Sedum album, Sedum spurium.

Previous page: Street-
Dome on Haderslev har-
bor front, moss sedum
roof. This page: rideable
terrain, interior climbing
wall, connecting inside
and outside.

The community center is located in the mountains of Taoyuan Park in Chongqing. The designers, with Dong Gong as the principal architect, sought to seamlessly integrate the new building outline into the existing undulating topography. Rather than simply building a structure in a field, they hoped to generate a fusion of architectural form and hilly landscape.

The green roof and walls help to blend the volume into its natural environment as well as enhancing the thermal efficiency of the building envelope. Interior and exterior spaces exist in harmony, complementing each other. The site's three main programs are the cultural center, athletics center and public health center. A continuous roof connects the

three independent buildings into one unified volume on a gross floor area of 10,000 square meters. Responding to the hilly site, the roof slopes up and down and frames two courtyards: a sloped garden and a green plaza where community activities take place. In traditional Chongqing architecture, Qilou (a style of arcade building) is a common feature due to the high level of rainfall. This principle has been adopted here within the outdoor circulation system, resulting in multiple paths connecting the two courtyards and building perimeter. The three major buildings each boast their own atrium where large skylights introduce natural light into the expansive space. Openings, windows, cantilevers and corridors blur the boundary between the indoor and outdoor areas, generating a lively coexistence of artificial structure and natural landscape.

Landscape design:
China Ctdi Engineering
Corporation.
Architecture: Vector
Architects. Structural
design: Congzhen Xiao.
Location: Chongqing,
China. Year: 2015.
Irrigation: drop irrigation
and automatic spraying
system.

Taoyuanju Community Center

Number of species: 43. Plants mainly used: Ficus virens, Ginkgo biloba, Zelkova serrata.

Previous page: green
plaza and athletic center.
This page: concrete cano-
py, entrance platform,
green courtyard.

King Street Live/Work/ Grow

Landscape design and architecture: Susan Fitzgerald. Structural design: Andrea Doncaster. Further participants: Servant Dunbrack McKenzie & MacDonald, Brainard Fitzgerald Developments. Location: Halifax, Nova Scotia, Canada. Year: 2014. Irrigation: as needed.

This mixed-use project is located within an eclectic community in the north end of Halifax, Nova Scotia. Its neighbors include the city crematorium, a print shop, a recycling depot, a coffee roaster and café, a car dealership, an automobile repair shop, condominiums, and the sparse remnants of row housing from the early 1900s. Within the 8 x 30 meter lot, the project contributes to the rich character of this lively community where the converging conditions of affordable land, rapid growth, and light industry suggest an uncertain future.

Advocating ways to maintain and enrich the multiplicity in the neighborhood, this project enhances density, community, and livability. Consisting of three separate units, each with entrances at grade, the program for the project includes an office space for an architecture and contractor firm, a dwelling for a family of four, and a two-story live/work rental studio apartment. Programmatic and spatial flexibility enables the commercial and residential spaces to contract or expand into one another based upon the viability of the business, and ever-changing family circumstances as children mature and parents age. Landscaped spaces are integrated throughout the whole project to offer respite within the city and support the cultivation of vegetables and flowers.

Number of species: 20. Plants mainly used: Ocimum basilicum.

Previous page: vegetables on lower roof, view of link with sleeping cubbies closed.
This page: view of two units looking east, King Street elevation, view from front unit roof, view of front unit from back unit.

Indoors

Vertical Garden Patrick Blanc

Botanist and vertical garden design: Patrick Blanc. Architecture: Mimolimit. Location: Bratislava, Slovakia. Year: 2010.

Do plants really need soil? No, they don't! Soil is nothing more than a mechanical support. Essential are water and an array of dissolved minerals, along with, of course, light and carbon dioxide for photosynthesis. When roots are allowed to penetrate a man-made wall, they easily damage and quicken its destruction. This peril can be avoided if vertical gardens are completely insulated from the existing wall. The green layer becomes the building's second, living, skin. Roots simply spread along the vertical surface of the support structure, leaving the inner wall unaffected; a harmony between plants and architecture is thus achieved.

Plants can be installed on this felt layer as seeds, cuttings or in already grown state. Irrigation is provided from above. If tap water is used, it must be supplemented with a low concentration of nutrients. Of course, the best solution is to recycle water, irrigating with either gray water, effluvient from adjacent roofs or condensation from air conditioning. The net weight of the vertical garden, including plants and the metal frame, is less than 30 kilograms per square meter. This makes it possible to be implemented on any wall, without limits as to height or size.

Thanks to its thermal insulation properties, the vertical garden improves efficiency and aids in lowering energy requirements, both in the winter (by shielding the building from the cold) and in the summer (by providing a natural cooling system). The vertical garden is also an efficient way to filter air. In addition to leaves and their well-known air-improving qualities, roots and the microorganisms associated with them act as a broad-range air cleaning ecosystem. The felt layer captures polluting particles and slowly decomposes and mineralizes them before utilizing them as a plant fertilizer. The vertical garden is thus an efficient tool for air and water remediation for instances when flat surfaces are already extensively used by human activity.

The vertical garden re-creates a living system very similar to natural environments. It is a way to reintroduce nature where it has been forcefully removed. Thanks to botanical knowledge and significant experience, it is now possible to create natural-looking plant landscapes. In any city, anywhere in the world, a bare wall can be transformed into a vertical garden, becoming a valuable shelter for biodiversity.

Number of species: 70. Plants mainly used:
Aeschynanthus Caroline, Ludisia discolor.

Previous page: view of
wall. This page: detail of
wall, reflection of seating
area.

Botanist

Landscape design and
architecture: Alberto
Caiola. Location: Shang-
hai, China. Year: 2016.
Irrigation: soil-free
hydroponic system.

Number of species: 10. Plants mainly used:
Clitoria ternatea, Rosmarinus officinalis.

Previous page: polycar-
bonate sheets create a
surreal perspective.
This page: exterior view,
the bar with large jars,
comfortable niches invite
to relax.

Botanist is a cocktail bar in Shanghai. Its architecture is a combination of natural and synthetic elements, forming a futuristic realm where technology meets sustainability to harness the power of plants. Seasonal herbs, spices, flowers, fruits and vegetables take center stage in the cocktail recipes, becoming the base for healthy and colorful drinks that respond to our growing desire for healthy, natural lifestyles. Upon entering the bar, the customer is presented with a floor-to-ceiling living wall of aromatic plants that immediately appeals to their olfactory sense. At the heart of Botanist's craft is the preparation of drinks, which happens on site and in full view thanks to a well-lit bar area. A wall installation featuring plants and polycarbonate sheets further intensifies the tension between natural and synthetic elements.

Depending on the viewer's perspective, the vegetation appears either abstract and blurred, or focused and sharp. To maximize the contrast, angular and synthetic components can be found throughout the space. Most dramatic of all is an immersive drop ceiling, a complex floating maze inspired by chemical structures. The irregular yet geometric frames resemble a synthetic forest where dark metal structures, strip lights, reflective panels and meshes come together to create semi-futuristic esthetics.

This page: view on the lush greened wall, geometrical shapes in the restrooms, the greenery breaking the strict geometrical shapes, view on the bar.

The joint venture KOOP and INNOCAD from Graz emerged as the winner in the 2010 competition to design Microsoft's headquarters in Vienna. Microsoft sets high standards for its buildings, aiming to create the ideal constellation of physical, social and virtual work environments for its employees. The designers of the renovated 4,500-square-meter headquarters in Vienna took the Work Place Advantage Concept, which analyzes the employee structure of every branch office, to heart and even extended it: the "sealed-off" employee floors were broken up and ingeniously arranged in a transparent manner. It is hoped that the painstaking attention to detail shown by both Microsoft and the designers will result in much higher levels of employee satisfaction, productivity and efficiency as well as a much reduced carbon footprint. An architectural "life-line" traverses the entire building in the form of accessible, multi-functional furniture, providing a spatial bracket around all of the floors and facilitating a variety of functional settings. The greatest possible flexibility is also provided in the closed meeting rooms. All high-traffic areas, such as corridors and foyers, were designed to be dynamic. The striped vinyl floor and a slide that allows quick access from the second to the first floor generate and inspire movement. Green walls are a prominent feature throughout the building, lending a unique atmosphere to the interior spaces. Lighting is provided by a small number of cleverly designed and positioned light sources. Neutral, uniform lighting in the form of linear elements in the "life-line" generates a calming atmosphere.

Landscape design:
Vertical Magic Garden.
Architecture: INNOCAD.
Location: Vienna,
Austria. Year: 2011.
Irrigation: integrated,
computer-controlled.

HQ Microsoft Vienna

Previous page: foyer
flooded with natural
light through vast glass
windows, open commu-
nication and access zone.
This page: each
conference room has
its own unique style,
the customer area and
cafeteria are ideal for
relaxation, floor plan.

Number of species: 15. Plants mainly used: Ficus pumila, Epipremnum aureum.

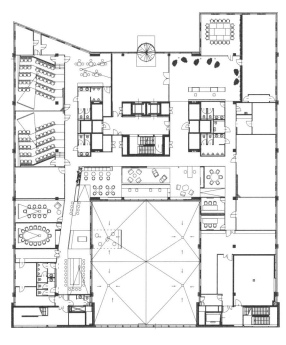

Stockholms-mässan, Galleries

Architecture: Alessandro Ripellino Arkitekter. Location: Älvsjö, Stockholm, Sweden. Year: 2010. Irrigation: central fertigation unit controlling the supply to all six walls, irrigated with dripline.

A-gallery and B-gallery form a communication path between the main entrance and the east entrance of this trade fair building in Stockholm. The galleries comprise two long rooms that differ in ceiling height, materials and uses. A-gallery is the higher and its ceiling is made of white expanded metal mesh. Behind the mesh the illumination is provided by diodes that change color on demand. B-gallery has a ceiling formed of reflective folded metal panels. The glossy reflection of the exhibition space that is reproduced multiple times by the panels evokes the idea of rippling water. Service functions such as rest rooms, vertical communications and the kitchen are housed between the two galleries and are clad with vertical green walls. The color helps to generate a strong relationship between these areas and the two exhibition spaces and to create a sense of the natural in contrast to the artificial large scale of the fair building. An astonishing 11,000 plants were used on a surface of around 250 square meters. which are irrigated via a dripline.

Number of species: 90. Plants mainly used: Philodendron, Begonia, Peperomia.

Previous page: reflective
panels above the green
wall. This page: stairs to
the second floor, details
of the greened wall, the
greened wall.

The Hang-ing Gardens of Berthelot

Interior design: I Love Colours Studio. Further participants: DelightFULL Unique Lamps. Location: Bucharest, Romania. Year: 2015. Irrigation: micro-drip system.

The Berthelot Hotel impresses the visitor with its beautiful interior garden. Forty flower pots made of powder coated pressed steel were filled with vegetation and now hang from the metal roof structure. A corner garden behind the sofa completes the picture of a green terrace. The organic approach featuring fluid curvaceous shapes is echoed in all details and interior fittings, contrasting with and complementing the rigor of the metal structure. Pleasant acoustics are guaranteed by wall padding and the addition of sound-absorbing materials. The image of raindrops on the glass surface of the roof inspired the idea of round mirrors, which are mounted on the curved wall above the sofa. As a result of natural light flooding the space, the covered terrace is the ideal location for breakfast or a lunchtime meal. Dinner however requires a more sophisticated ambience, which is enabled through DelightFULL's wall lamps with their long adjustable copper arms. The scissor arms of the Pastorius Wall Lamp allow the guests to direct the light as they wish.

Number of species: 12. Plants mainly used: Sansevieria trifasciata, Spathiphyllum.

Previous page: the tables
reside beneath lush
greenery. This page:
natural light illuminates
the spacious restaurant,
the adjustable lamps are
highly convenient.

Home
06

Landscape architect:
Green Fortune.
Architecture: i29 interior
architects. Constructor:
M. J. Rasch. Interior
build: H2B interiors.
Location: Amsterdam,
the Netherlands.
Year: 2012. Irrigation:
drip irrigation tubes.

This residence at the Singel, Amsterdam, exists as one open space where freestanding objects have been attributed several functions. For example, the kitchen and wardrobe are placed near the entrance and combined into a single volume. The combined bed-/bathroom is hidden away in a volume placed at the rear of the house, while the open living area offers views of the vertical garden and entrance stairs leading to the roof terrace. The view towards the green wall tantalizes the observer, withholding its secrets until one enters the bed-/bathroom.

The compact size of this space contrasts with the overall spaciousness of the residence and generates a welcome atmosphere of privacy and intimacy. The juxtaposition of the overgrown plant wall and the minimalist white bed-/bathroom offers a second contrast that enhances the esthetic pleasure of each element. Integration of nature is an important aspect in the traditional culture of Japan, the native country of the client. This principle finds perfect expression in the in-house vertical garden, a cornucopia of lush green vegetation.

Number of species: 8. Plants mainly used: Philodendron.

Previous page: side view on the greened wall, room concept.
This page: view on the stairs, partition between the rooms, view on the shower, the spacious and light kitchen.

Hotel Kungsträd- gården

Landscape design: Vertical Garden Design. Architecture: Link Arkitektur. Location: Stockholm, Sweden. Year: 2014. Irrigation: automated with dripline and moisture sensor.

Hotel Kungsträdgården in central Stockholm is located in a building dating back to the 18th century. An interior patio with a glass ceiling houses the restaurant and bar. The two vertical gardens are narrow but tall, one covering the wall from floor to ceiling, an expanse of 15 meters. The plants are arranged in a vertical pattern, which will generate the impression of cascading vegetation once the plants reach a mature state. Larger group plantings of Philodendron scandens, Ficus pumila, Aeschynantus, Scindapsus, Asparagus and others form a pleasant backdrop for the accent plants. Several kinds of aroids, including Philodendron giganteum and Philodendron "Imperial Green" were selected as accent plants. Most of these will develop large pendant leaves over time. Some have cascading growth habits and others tend to develop gradually larger leaves as the plant reaches its mature form. Begonia and Billbergia offer flowering accents, while the long and cascading ferns of Polypodium and Nephrolepis enhance the wall's floral diversity.

Previous page: lounge
area on the third floor.
This page: view on the
greened walls and the
luxurious lounge, view
on the restaurant.

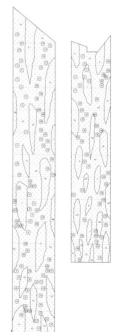

Number of species: 54.
Plants mainly used:
Scindapsus pictus.

This page: details of the greenery, the lush green complements the elegant ambience of the restaurant, overlooking the chic and spacious restaurant.

Beko IFA 2014

Architecture: einszu33.
Green decoration: Folke
Frey of Liebesdienste.
Location: IFA Berlin,
Germany. Year: 2014.
Irrigation: manually.

The fourth trade fair appearance of the firm Beko at IFA 2014 in Berlin revealed a focus on global responsibility and a close connection with the customer. It was a public park in New York that inspired this vision. The High Line was created by the city's residents on a section of old railway. At Beko's stand visitors could experience an urban landscape informed by a fusion of architecture and urban gardening. They were also greeted with a greened roof terrace, complete with beehives, a live cooking station, a sustainability zone, an urban gardening area filled with lush vegetation, an award zone and an "inspiration campus", inspired by a social art project in Brazil. A recuperation zone formed the middle part of Beko's exhibit. Three real pine trees and the sound of twittering birds conveyed the feeling of sitting in the midst of nature and generated an atmosphere of relaxation and tranquility.

"GOOD IDEAS ARE LIKE GOOD PLANTS – THEY CAN GROW ANYWHERE." INSPIRED BY NATURE LOVERS.

Previous page: crop plants arranged appealingly. This page: herbs and vegetables, saving nature, the urban gardening area invites visitors to discover their green thumbs.

Number of species: more than 1,000.
Plants mainly used: Hedera helix, Pinus.

The architectural landscape in many rapidly developing cities in Asia is becoming increasingly uniform. These urban areas are losing their regional characteristics under the influence of furious urban sprawl and commercialization, while rapidly growing populations are causing a decline in both quality of life and green spaces. Ho Chi Minh City, the largest city in Vietnam, is no exception. Stacking Green, a prototypical private house built in 2011, is an attempt to confront and redirect this trend. The city's residents love living with plants and flowers in their streets. Even in this modernized urbanized space, many city dwellers unconsciously desire the experience of a tropical forest. Stacking Green realized this desire with a façade composed of planters like horizontal louvers. This green façade not only contributes to the project's esthetic qualities, but also upgrades the indoor thermal environment, thus saving energy. The semi-open green screen also increases privacy. The house is a typical tube house constructed on a plot of 80 square meters. The front and rear façades are composed of layers of concrete planters cantilevered from two sidewalls. The green façade and roof garden protect the inhabitants from direct sunlight, street noise, and pollution. According to post-occupancy measurements of the indoor environment, wind flows through the house thanks to the porous façades and two skylights. This is borne out by the minimal usage of the air conditioning system, remarkable in this tropical climate.

Landscape design and architecture: Vo Trong Nghia Architects.
Main architects: Vo Trong Nghia, Daisuke Sanuki, Shunri Nishizawa.
Contractor: Wind and Water House JSC.
Location: Ho Chi Minh City, Vietnam. Year: 2011.
Irrigation: automatic and with rain water.

Stacking Green

Previous page: green,
concrete and wood are
combined harmoniously
in this project. This page:
view on the rooftop
garden, exterior view,
detail of greenery.

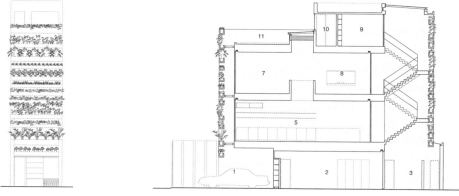

Number of species: 7. Plants mainly
used: Saribus rotundifolius.

This page: natural light permeates the lush greenery, the green façade enhances the light and climate of the interior, the skylight in the bathroom floods the bathing area with light.

Parkroyal on Pickering

Landscape design: Tierra Design (S) Pte. Ltd. Architecture and interior design: WOHA. Location: Singapore. Year: 2013. Irrigation: automatic.

Fronting Singapore's Hong Lim Park, Parkroyal on Pickering, a Platinum-Greenmark hotel and office development demonstrates that increased density can come with an increase in amenity. The building form contrasts modernist or-thogonal forms with a romantic abstract-ed contoured landscape. The contour expression creates a series of planted terraces on the upper surfaces, and Chinese-grotto inspired spaces below. The top of the podium is a romantic landscape that covers the entire site, and features "Birdcage" pavilions that hang out over the street. The room blocks, organized as an E shape, support hanging gardens between them, giving foreground interest to the views over the park. The room blocks have detached, naturally ventilated corridors facing the public housing blocks behind, these are flanked by water and gardens on the exterior side, and a green wall against the rooms. Finally, at the top of the building, planted roof terraces give great views over the city.

Previous page: looking up
to the underside of the
sculpted sky terraces.
This page: swimming pool
and cabanas, view of the
urban verandah, view
from the street.

Number of species: 50. Plants
mainly used: Vernonia elliptica.

This page: view of the sky terraces from the club lounge, a naturally ventilated guestroom corridor, the lift lobby looking towards the all-day dining restaurant, 14th floor plan.

Elok House was conceived as a three-dimensional landscaped installation into which the living spaces were inserted. Floating platforms were generated that optimize natural day lighting, cross-ventilation and views. The living spaces are defined by voids and natural landscape elements such as plants, pebbles and water. In the construction of Elok House, a conventional intermediate terrace house was reconfigured with a central zone, flanked by one-meter wide linear light and ventilation wells along both sides of the site. The ground floor is one interconnected landscape, with the kitchen located at the front as part of the entrance porch setting. The old boundary wall was removed, pushed back a short distance and replaced with thin lines of steel cables held by sliding frames. These transformations have generated a cordial setting in which meeting visitors is now a pleasurable experience. Beyond the entrance, the casual/formal dining and living areas are interchangeable, defined loosely by landscape elements. The space culminates at the rear with an existing retaining wall spectacularly transformed into a waterfall, flanked on both sides by vertical green walls. The configurations of the spaces, in combination with the luxuriant use of plants and water elements, generate a cool microclimate within the house, which is also pleasantly bright and airy during the day. The completed interior resembles a tropical rainforest with minimal need for air conditioning or artificial lighting, a sublime living environment where one can be in sync with nature in the midst of urbanized life.

Landscape design and
architecture: CHANG
Architects. Further
participants: DPC Con-
sulting Engineers, Tim
Contracts, Greenscape.
Location: Jalan Elok,
Singapore. Year: 2008.
Irrigation: automatic
irrigation system.

Elok
House

Previous page: the living
room – a lush oasis.
This page: bedroom with
greened wall, exterior
view on the townhouse,
detail of plants.

Number of species: 20. Plants mainly used: Asplenium nidus.

This page: the terrace resembles an urban version of the Garden of Eden, view on the greened stairs, kitchen with trees and natural lighting.

Index

Picture Credits

Cover front / left (from above to below): Patrick Blanc, Hufton + Crow, Paolo Rosselli, Kengo Kuma & Associates
Cover front / right (from above to below):
Paul Bardagjy, 90deGREEN
Cover back (from above to below, from left to right):
Optigrün international, Dirk Weiblen, Paul Ott